IBM. (2022). *AI task automation frameworks for enterprise scalability*. IBM Research White Paper.

Lucas, H. C., & Goh, J. M. (2009). Disruptive technology: How Kodak missed the digital photography revolution. *Journal of Strategic Information Systems*, 18(1), 46–55.

Norman, D. A. (2013). *The design of everyday things* (Revised and expanded ed.). Basic Books.

VentureBeat AI Weekly. (2023). *Fragmentation in the AI tooling market*. Retrieved from https://venturebeat.com

Wallach, W., & Allen, C. (2008). *Moral machines: Teaching robots right from wrong*. Oxford University Press.

Chapter 7: Powered by AI Freelance — Building the AI Task Universe

Airbnb, Inc. (2024). *How Airbnb works*. Retrieved from https://www.airbnb.com/help

Amazon.com, Inc. (2024). *Amazon Marketplace Services*. Retrieved from https://services.amazon.com

Apple Inc. (2008). *The App Store launch announcement*. Retrieved from https://apple.com/newsroom

Auto-GPT. (2024). *GitHub repository*. Retrieved from https://github.com/Torantulino/Auto-GPT

CrewAI. (2024). *Open agent orchestration platform*. Retrieved from https://github.com/joaomdmoura/crewai

Fiverr. (2023). *Annual report*. Retrieved from https://investors.fiverr.com

Google. (2023). *Scaling TPU compute and cloud networking*. Alphabet earnings call transcript.

Hugging Face. (2024). *Company overview and model access.* Retrieved from https://huggingface.co

LangChain. (2024). *Documentation and overview.* Retrieved from https://docs.langchain.com

McKinsey & Company. (2023). *The gig economy and digital labor platforms.* Industry insights and projections.

Shopify. (2024). *Partner and app marketplace overview.* Retrieved from https://www.shopify.com/partners

Uber Technologies Inc. (2024). *How Uber works.* Retrieved from https://www.uber.com

Upwork. (2023). *Freelance Forward: 2023 report.* Retrieved from https://www.upwork.com/research/freelance-forward-2023.

U.S. Census Bureau. (2023). *Freelance and independent worker trends in the U.S. economy.* Referenced economic data and projections.

Chapter 8: Securing the Future — Data Control, Compliance, and Trust

American Institute of CPAs (AICPA). (n.d.). *Trust Services Criteria for SOC 2.* Retrieved from https://www.aicpa.org

Ammanath, B. (2022). *Trustworthy AI: A business guide for navigating trust and ethics in AI.* Wiley.

Federal Trade Commission (FTC). (2023). *Consumer protection and data breach enforcement record.* Referenced in the Finwell Capital case study.

Historical business case: Borders Group and Walmart, derived from publicly available financial and press sources.

Isaac, M. (2018). *Facebook's role in Cambridge Analytica data scandal.* The New York Times.

Krach, K. (2020). *Public speeches and published materials on the Trust Principle and digital trust diplomacy*. See https://www.keithkrach.com

Ryan, J., & Lazo, M. (2023). *AI data privacy and protection: Managing risk in the age of machine learning*. TechPress.

U.S. Census Bureau. (2024). *Data on AI adoption among U.S. businesses*. Referenced statistics.

Chapter 9: Redefining Work

Daugherty, P., & Wilson, H. J. (2018). *Humans + machines: Reimagining work in the age of AI*. Harvard Business Review Press.

Leonardi, P., & Neeley, T. (2022). *The digital mindset: What it really takes to thrive in the age of data, algorithms, and AI*. Harvard Business Review Press.

Susskind, D. (2020). *A world without work: Technology, automation, and how we should respond*. Metropolitan Books.

Eloundou, T., Manning, S., Mishkin, P., & Rock, D. (2023). *GPTs are GPTs: An early look at the labor market impact potential of large language models*. OpenAI.

World Economic Forum. (2023). *Future of jobs report*. Geneva: WEF.

Goldman Sachs. (2023). *The potentially large effects of artificial intelligence on economic growth*. Goldman Sachs Research.

McKinsey & Company. (2024). *The next frontier of customer engagement: AI-enabled customer service*. Retrieved from https://www.mckinsey.com

npj Digital Medicine. (2024). *AI diagnostic accuracy across specialties: A systematic review*. Springer Nature.

ScienceDirect. (2025). *Generative AI and the creative professions: Risk or reward?* Elsevier.

Historical reference: Richard Arkwright and the invention of the water frame during the Industrial Revolution, c. 1769.

Historical reference: The digital transformation of the workplace in the 1990s and early 2000s with the rise of the internet, email, and early CRM platforms.

Chapter 10: Building the AI-Augmented Workforce

Amazon. (2023). *Career Choice: Upskilling our hourly employees*. Company press release.
Cited for internal workforce mobility and certification parallels.

Anthony, S. D. (2016). *Kodak's downfall wasn't about technology*. Harvard Business Review. https://hbr.org/2016/07/kodaks-downfall-wasnt-about-technology

Daugherty, P., & Wilson, H. J. (2018). *Humans + machines: Reimagining work in the age of AI*. Harvard Business Review Press.

Lamarre, E., & Smaje, K. (2023). *Rewired: The McKinsey guide to outcompeting in the age of digital and AI*. Wiley.

Pink, D. H. (2009). *Drive: The surprising truth about what motivates us*. Riverhead Books.

PwC. (2022). *How our digital accelerator program is upskilling future-ready consultants*. PwC Newsroom.
Supports Task Workforce and internal upskilling alignment.

Smith, J. (2021). *AI in business*. Future Insights Publishing.

West, D. M. (2018). *The future of work: Robots, AI, and automation*. Brookings Institution Press.

Historical reference: Richard Arkwright's water frame, c. 1769.

Historical reference: IBM's workforce reinvention and digital transformation, 2000s.

Historical reference: Kodak's decline following digital disruption, 2000–2012.

Quote: Nadella, S. (2023). *Microsoft CEO keynote on leadership and AI.*

Quote: Rometty, G. (2020). *World Economic Forum address on workforce transformation.*

Chapter 11: The AI-Powered Business Playbook

Anderson, J. L. (2021). *Artificial intelligence for business: A roadmap for profitable and responsible AI adoption.* Future Insights Press.

Bennett, M. (2023). *A brief history of intelligence: Evolution, AI, and the five breakthroughs that made our brains.* Ballantine Books.

Grant, A. (2021). *Think again: The power of knowing what you don't know.* Viking.

Horowitz, B. (2014). *The hard thing about hard things: Building a business when there are no easy answers.* Harper Business.

Kotter, J. P., & Akhtar, V. (2021). *Change: How organizations achieve hard-to-imagine results in uncertain times.* Wiley.

LIBBi platform (as proprietary model). Taken from http://www.libbi.co

Sweetgreen: Case study based on public interviews and operational insights published in *Fast Company, Inc.,* and industry press (2019–2023).

Chick-fil-A: Digital operations model referenced from company disclosures, interviews, and industry analysis (2020–2023).

Chapter 12: Leadership Revisited — Change

Blockbuster and Netflix: Business history synthesized from interviews and retrospectives published in *Forbes, Harvard Business Review,* and *The New York Times.*

Brown, B. (2018). *Dare to lead: Brave work. Tough conversations. Whole hearts.* Random House.

Fireclay Tile: Business transformation under CEO Eric Edelson, covered in *Inc. Magazine* and *Medium*.

Grant, A. (2021). *Think again: The power of knowing what you don't know*. Viking.

Home Depot: Strategic evolution under Craig Menear and Ted Decker. Public reporting and press interviews (2014–2023).

Hutton, T. (2023). *Leadership philosophy and cultural strategy*. Personal correspondence; former IBM executive and investor.

Kouzes, J. M., & Posner, B. Z. (2017). *The leadership challenge: How to make extraordinary things happen in organizations* (6th ed.). Wiley.

Kotter, J. P. (1996). *Leading change*. Harvard Business Review Press.

Kotter, J. P., & Akhtar, V. (2021). *Change: How organizations achieve hard-to-imagine results in uncertain times*. Wiley.

Langone, K. (2018). *I love capitalism!: An American story*. Portfolio.

Nadella, S. (2023). *Remarks on AI leadership and organizational empowerment*. Public keynote; referenced in Microsoft corporate communications.

Rometty, G. (2020). *Remarks at the World Economic Forum on leadership in workforce transition*. WEF Annual Meeting.

Union Kitchen: COVID-era pivot and operational model referenced in *Washington Business Journal* and local business press.

Appendix 1: Make the Move to AI Playbook

This guide is designed to help you create a playbook to effectively move your business into the era of AI. You will identify strengths and weaknesses of your current operation. Importantly, you do not have to complete this guide to act now and make the move. But doing so will help you organize your thoughts, prioritize what is most important, quantify the benefits of making the move, and more effectively manage the change in your organization.

An electronic version of this playbook is available on our publisher's site: www.intelligentpress.co. We, also, would love to hear from you–share your playbook, additional questions you add, feedback about the book, new AI tasks you imagine. Reach us at support@intelligentpress.co (please add *Navigating AI for Business* to the subject line).

Step 1: Assess Your Business Workflows

What are your core workflows—and which ones should be automated first?

Identify the key operational workflows or processes your business relies on, then rate each by its current ease of use (1 = easy, 5 = very difficult) and importance (1 = low priority, 5 = mission-critical). This helps determine what Tasks to activate first.

Workflow	Difficulty to Use (1–5)	Importance to Scale (1–5)
Prospect Identification		
Bidding & Proposal Generation		
Service Tracking & Fulfillment		
Customer Onboarding		
Marketing Automation		

Social Media Management

Appointment Scheduling

Financial Reporting & Invoicing

Document Tracking

Materials Management

_____

_____

_____

Step 2: Define the AI Experience You Want to Deliver

What experience do you want to deliver to customers, employees, and partners?

Who do I want to benefit from our new AI-powered experience?

> Employees of our business
> Customers of our business
> Potential customers of our business

What Do You Want Your Stakeholders to Be Able to Do With Your New AI Tools?

Core Business Interactions

Complete full workflows or submit requests end-to-end

Book appointments, consultations, or services

Purchase products, packages, or subscriptions

Pay invoices, deposits, or recurring billing online

Submit forms, applications, or required documents

Approve proposals, estimates, or change orders

Track project, job, or service status in real time

Receive automated updates, reminders, or alerts

Communication & Support

Chat with an AI assistant for common inquiries

Submit tickets or requests for support or service

Receive guided onboarding or product walkthroughs

Schedule calls or follow-ups with team members

Leave feedback or complete satisfaction surveys

Data Access & Content Delivery

Access historical documents, invoices, or reports

Download personalized content (e.g., user guides, deliverables, dashboards)

View and manage account or subscription details

Get AI-generated summaries, analytics, or recommendations

Business & Performance Insights

Monitor progress or task completion milestones

Receive AI-driven reports or diagnostics (e.g., financial performance, usage trends)

Interact with dashboards powered by unified data

Run what-if scenarios or forecasting Tasks

Receive AI-driven analysis and recommendations

Permissions & Workflow Approvals

Grant access or set permissions for internal Tasks

Approve expenses, budgets, or personnel requests

Review and sign digital contracts or agreements

Configure preferences or notification settings

Partner & Vendor Functions

Submit invoices or request payment

Respond to bid requests or project solicitations

Share compliance documentation or certifications

Coordinate schedules and deliveries

Additional:

_____

_____

_____

_____

Step 3: Evaluate Your Data Sources

Where is your data currently stored—and is it secure and centralized?

Where is your business data currently stored?

☐ Local drives or external hard drives

- ☐ Google Drive / Dropbox / iCloud
- ☐ CRM platform (e.g., HubSpot, Salesforce)
- ☐ Accounting system (e.g., QuickBooks)
- ☐ Project management platform (e.g., Asana, Trello, Notion)
- ☐ Email or text
- ☐ Printed documents / files
- ☐ Other: _____

Is there a central 'source of truth' for your customer and task data?

☐ Yes   ☐ No

Is your data and customer information secure?

☐ Yes   ☐ No

Step 4: Identify Key People to Engage

Who needs to be involved in this transformation?

Who are the team members or collaborators that should be active participants in making the move to AI?

Name	Role / Function	Reason for Involvement

Step 5: Platforms I'm Currently Using

What platforms are you already paying for?

Use this section to identify all SaaS platforms your business currently pays for. Estimate your monthly cost and list the function of each platform. This will help you determine your current business spend on software and what efficiencies you can gain by replacing with a unified platform or other solutions.

Platform Inventory:

Platform Name	Estimated Monthly Cost	Primary Function

Step 6: Select Business Operating Policies to Adopt

Which business policies do you need to adopt or update?

When making the move to AI it's important to ensure your business aligns with best practices for data governance and legal compliance. Select which policies you'd like to adopt or review during implementation:

- ☐ Privacy Policy
- ☐ Terms & Conditions of Use
- ☐ Cookie Policy
- ☐ GDPR Compliance Policy
- ☐ Data Security and Retention Policy
- ☐ AI Usage and Transparency Disclosure
- ☐ Other: _____

Step 7: Additional Support

What additional support do you need?

AI transformation is not a solo sport. It requires smart planning, thoughtful execution, and the right people—at the right time—to bring it all together. This section invites you to reflect on where your business could benefit from expert support.

What additional resources do you need to support your move to AI or to create new efficiencies in your business operation?

☐ Do I need a Task Generalist to help implement workflows?
☐ Could a Task Specialist help accelerate value creation?

Step 8: Plan for Growth

How will you measure and manage growth?

Transformation without direction is chaos. That's why this section prompts you to define clear, achievable goals—typically in 30/60/90-day increments—so you can track your progress, adjust in real time, and build sustainable momentum. You'll also outline how your team will review what's working, identify what needs improvement.

Define growth goals (tied to move to AI):

Goal:	30/60/90 days (identify):	Who Owns it:

Define success metrics:

Metric:	30/60/90 days (identify)	How Measured?

Step 9: Build My Own

What new Tasks would accelerate your business?

This final section invites you to look beyond what's available today and puts you in the position of creating AI for tomorrow. As you assess your workflows, stakeholder needs, and industry challenges, consider where a custom AI Task—tailored to your exact process—could unlock a new level of performance, automation, or customer engagement.

This is your opportunity to identify the gaps—whether it's a task you've never seen before, a workflow unique to your niche, or a system you've always wanted to automate but couldn't find a tool that fit.

Describe, in natural language, what you wish you could accomplish using your AI-powered business solution:

If you could build an AI Task to accomplish this work, what impact would it have on your business?